Collins *gem*

Origami

Trevor Bounford

Harper CollinsPublishers
1 London on Bridge Street, London SE1 9GF

www_collins.co.uk

A Diagram book first created by Diagram Visual
Information Limited of 195 Kentish Town Road,
London NW5 2JU

First published 2000
This edition published 2005

19
Reprint 12

© Diagram Visual Information Limited 2000, 2005

ISBN-13 978-0-00-718881-9

Printed in China

Introduction

Origami is a Japanese word which literally means 'paper folding'. The art of paper-folding originated in Japan many centuries ago, but it now has a growing popularity worldwide. As a pastime or hobby it has a universal appeal, whereby children and adults alike can feel the magic of creating attractive, and useful, objects out of something as simple as a sheet of paper.

With a little time, patience and practise you can easily develop the skills needed to make the models illustrated in this book. Very few tools are needed and the material costs are small, unless you choose to use expensive handmade papers. And there is virtually no cleaning up when you finish. It is also an ideal holiday craft – all you need are a few sheets of paper and you can while away rainy hours creating beautiful items.

This book includes a number of models varying in difficulty from very simple to fairly complex. However, with care and patience, and following the clear instructions and the simple drawings, anyone should be able to produce these origami artefacts. As with most skills, practise is the key. Practise making the different types of fold using scrap paper before attempting a real model, and, particularly when trying one of the the more complex items, rehearse the whole process with a scrap sheet so you are familiar with every step. That way you will be able to enjoy every part of this challenging and rewarding practical craft.

Contents

Decorative origami

Paper toys

Stationery items

Materials, tools and method

One of the benefits of origami is the relative low cost of the materials required. At least to start with you can use inexpensive paper such as photocopier paper – which comes in various colours – or any other thin, strong paper. Old telephone directory pages or even magazines and newspapers can be used for practising models. In fact newspaper is very good for hats, especially if you need to improvise a sun hat and no other material is available. Paper with a printed grid is also useful for practise as the lines will help you to fold accurately.

Special origami paper, which is coloured on one side only is, of course, ideal as it is designed to fold well. Experiment with other papers. Bright or pastel colours produce attractive origami models. Some textured papers or even reflective foil papers can be used, as long as they crease well. You can try gift wrap or other patterned papers for interesting effects. Just a few examples of the variety of papers that can be used are shown on the following pages. The important thing to bear in mind is that the paper must fold cleanly, leaving a sharp crease, and that the colour or texture doesn't crack when it is folded.

Most important of all, do practise with disposable paper, such as newsprint or before you attempt to produce any origami model using expensive paper.

PAPER

Origami paper is specially designed to fold cleanly and usually has colour on one side only.

Brightly coloured papers are good for toys and some of the animal models.

Subtle pastel-coloured papers make tasteful origami.

Newspapers and old telephone directory pages are good for practising. Gridded paper will help to develop accurate folding technique.

Metallic-coated foil paper can make exciting models but make sure it folds well and doesn't crumple easily.

Textured papers are also good for origami as long as the texture does not prevent clean folds and the paper is not too thick.

Gift wrapping paper can be used to produce bright and interesting effects.

TOOLS

Useful tools for origami include: **1** a sharp craft knife (which must be handled with great care); and **2** a pair of good scissors; **3** a well-sharpened pencil or **4** a draughting pencil, for marking measurements; **5** an artist's kneadable eraser for removing marks without damaging the paper; **6** a metal ruler for measuring and cutting, and also for using to hold parts of models flat when making difficult folds; **7** a set-square; and **8** a self-healing cutting mat – ideally one with a grid.

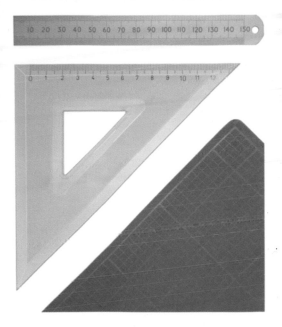

METHOD

The art of successful origami lies in the technique of folding. Although anyone can fold a sheet of paper, making accurate folds cleanly and consistently requires some practise. The initial folds of even a simple origami model are crucial to the successful completion so it will pay to follow the few simple guidelines set out here.

Make sure your hands are clean – a fine model will be marred by grubby marks.

Use a clean, flat, firm and level surface set at the right height to work on – a table-top or a drawing board in your lap. Keep the area clean and have some suitable receptacle for scrap close at hand.

Make sure the corners of the paper are square – it is impossible to produce accurate folds if the paper is not a true rectangle to start with.

Practise making precision folds with scrap paper. Don't start a complex model with your best piece of paper without having first rehearsed all the stages on discardable paper.

In this book many of the instructional illustrations have had to be placed so they best fit on the pages – they are not necessarily the ideal orientation for the fold described. Although there are no set rules for making folds some established techniques should be helpful. Most people find that it is easier, and more accurate, to make the first few folds by lifting the nearest edge or corner and folding that to the edge or corner away from the body. Then rotate the folded sheet so that you can make the next fold in a similar direction. As the model develops you will find it more efficient to hold it in a particular position and fold in whichever direction is necessary for that step.

To make a fold **1** align the corners and edges precisely, **2** hold them firmly together then run one finger from the held edge or corner down to form a crease at the centre of the fold line. Now, without releasing the gripping point, run the finger out from the centre to one edge **3** and then back and across **4** to the opposite edge. Check that the aligned corners and edges are still in place and that the fold is flat and complete.

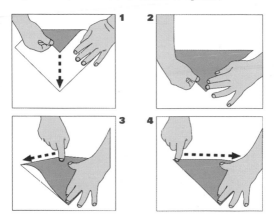

KEY TO FOLDS

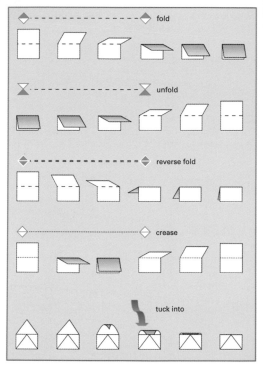

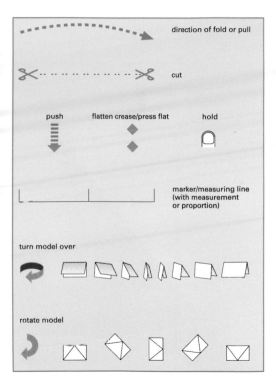

direction of fold or pull

cut

push   flatten crease/press flat   hold

marker/measuring line
(with measurement
or proportion)

turn model over

rotate model

Simple projects

CLAPPER

You will need

an oblong piece of paper about 40 cm x 20 cm to make this simple noise-maker. You can use a fairly stiff paper as there are only a few folds to make.

1 Fold and unfold to crease the sheet down the middle of the longer dimension. **2** Fold corner **A** so that the top edge lies down the centre line.

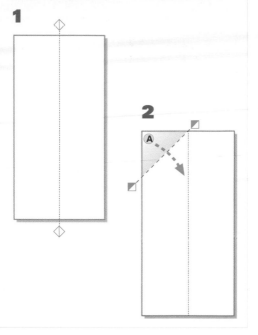

3 Now fold down corner **B** so its edge lies on the centre line. **4** Similarly fold in corner **C**.

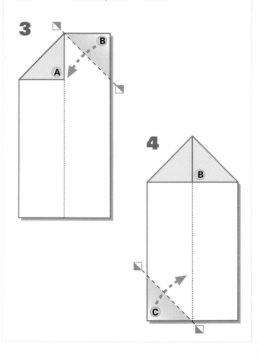

5 Fold in corner **D** so its edge too lies along the centre crease. 6 Now fold the sheet in half so that corner **E** lies on **F**.

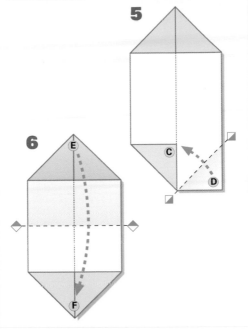

7 Fold edge **G** across the centre line. **8** Holding down all the underlying model pull the top flap at **G** up and over so that **H** comes to the centre **9**.

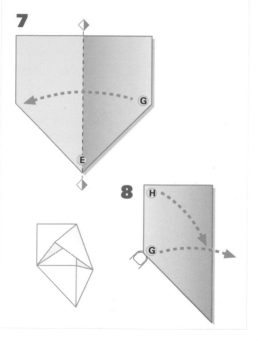

10 Leaving **H** where it is, fold **G** back over. Turn the model over.

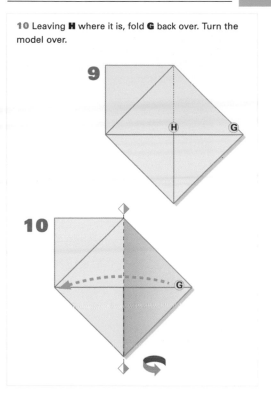

11 Holding down the underlying flaps, pull **I** up and over so that **J** falls in the centre **12**. **13** Now fold **I** back.

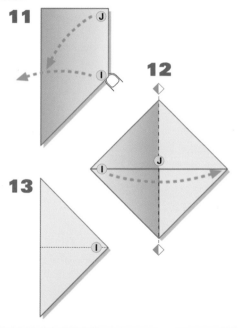

To operate the clapper, hold the bottom corner tightly and bring the clapper down sharply in an arc. This opens the inner flaps and claps them together. Push the flaps back in to repeat the action.

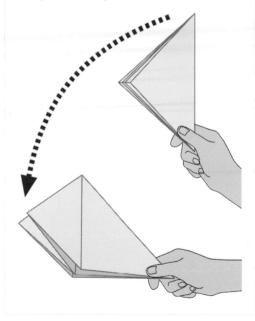

SIMPLE HAT

You will need

an oblong piece of paper about 60 cm x 40 cm to make a hat that will fit a child's head. Use a double page from a tabloid newspaper, or a single page from a broadsheet. This is a a good starter project but the hat can be very useful as an 'emergency' sun shade.

1 Make a centre crease lengthways along the sheet.
2 Fold edge **A** to the opposite edge then rotate the sheet so **A** is at the base.

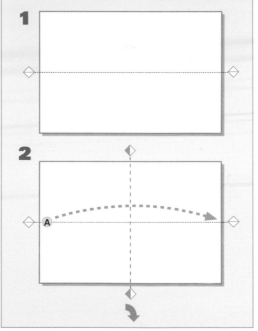

3 Fold corner **B** so that the top edge runs down the centre line. **4** Fold corner **C** to do the same.

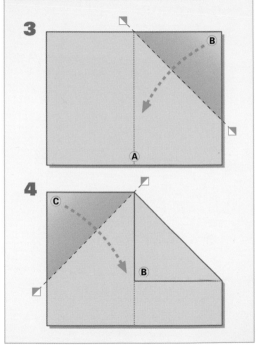

5 Fold edge **A**, upper sheet only, so that it touches the base of the triangles. **6** Now fold flap **A** again so that it overlaps the bases of the triangles.

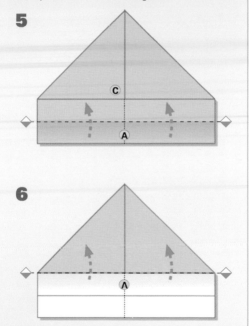

7 Turn the model over and **8** fold up edge **D** to the base of the triangles.

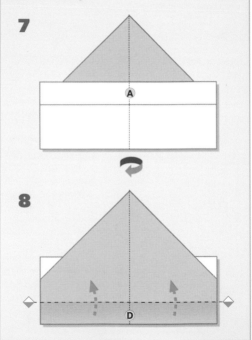

9 Fold up flap **D** once again to overlap the bases of the triangles. You should now have a triangular shape with two folded flaps and an opening at the base **10**.

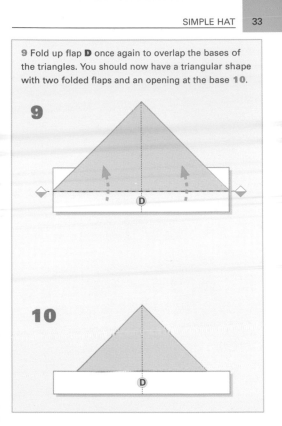

Pull apart the flaps at the bottom to open up your hat!

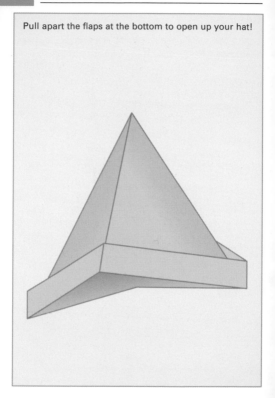

BASIC BOX

You will need

an oblong piece of paper of proportion 2:3. You can vary the actual size to produce different sized boxes but a piece measuring 20 cm x 30 cm would be sensible to start with.

1 Divide the length of the paper into three and fold edge **A** at the first mark so it touches the second. **2** Fold it back so it lies along the first fold line.

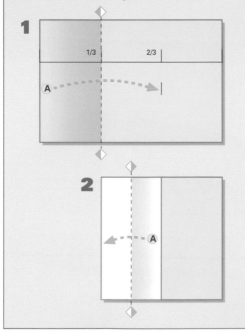

3 Fold edge **B** to lie over **A**, then fold it back again to the right-hand fold 4. 5 Unfold **B** to lie on **A**.

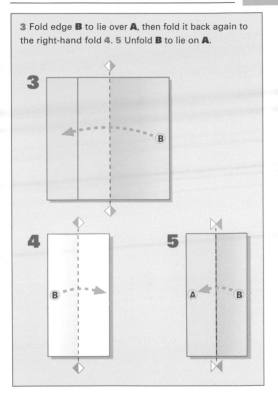

6 Now rotate the model. **7** Fold corner **C** to lie along the central crease – only the upper layer is folded.

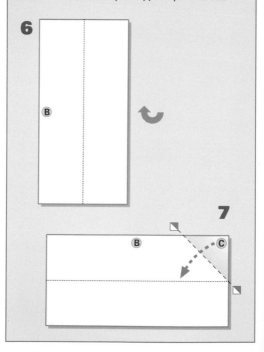

8 Similarly fold the upper layer of corner **D** so that it lies along the centre crease. **9** Now fold up corner **E** (which is a double thickness).

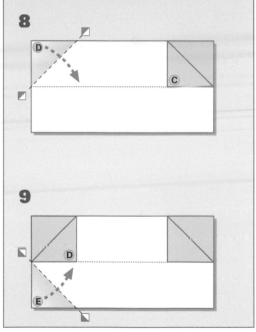

10 Fold corner **F** (also a double thickness) in the same way to lie along the centre crease. **11** Fold flap **B** down along the centre crease.

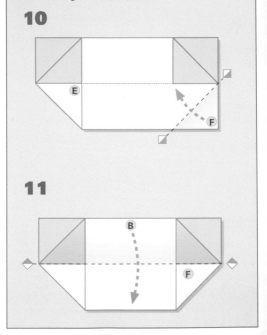

12 Now fold flap **A** (a single thickness) down to lie over flap **B**. **13** Fold corner **G** (a single thickness) up to lie along the centre crease.

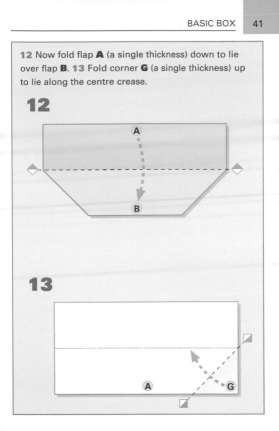

14 Next fold corner **H** (also a single thickness) in the same way to lie along the centre crease. **15** Fold down corner **I** (double thickness).

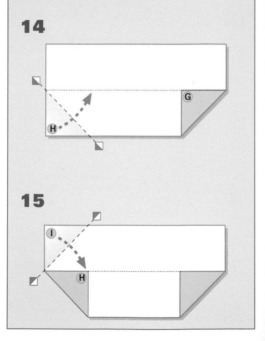

16 Next fold down corner **J** to the crease. **17** Fold flap **A** back up again.

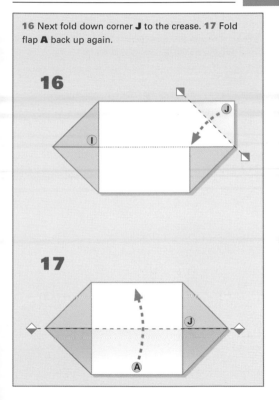

18 Carefully open the central slit and pull the two sides, **K**, apart. **19** Once the box has taken shape you will need to crease the edges to finish it.

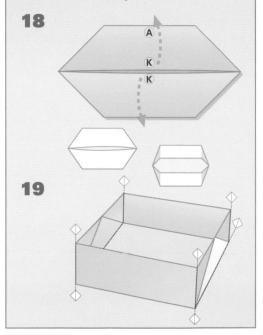

SWAN

You will need

a square piece of paper 20 cm x 20 cm to make a swan
about 10 cm high. White or pale cream paper would be
most swanlike – a stronger coloured paper has been
used here to help show the method.

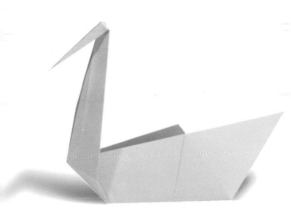

1 Crease the paper diagonally. **2** Fold corner **A** so the edge lies along the centre crease.

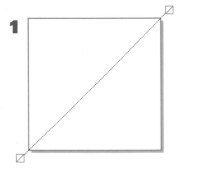

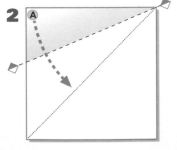

3 Fold up corner **B** so that edge also lies along the centre crease. **4** Now turn the model over.

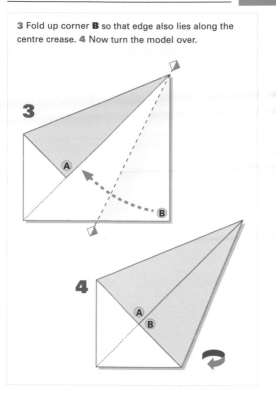

5 Fold corner **C** to lie along the centre crease. **6** Fold corner **D** to do the same.

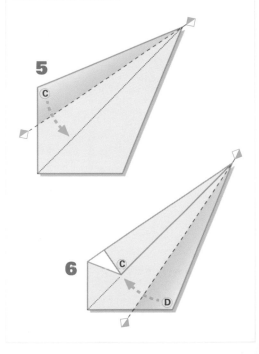

7 Fold corner **E** over to lie on corner **F**. **8** Measure the folded length and divide into three units.

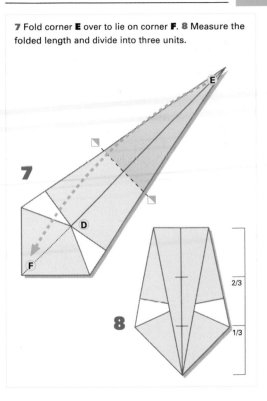

9 Fold back corner **E** at the first third. **10** Now reverse fold the model along the centre line.

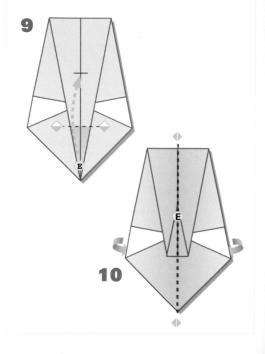

11 Holding the model as shown, carefully pull up the head and neck. **12** Flatten the new fold at the front of the body.

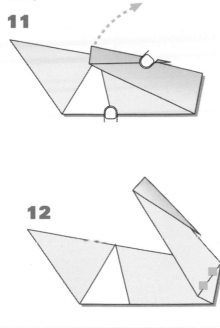

13 Now pull the head up slightly – swans tend to look down their beaks – and flatten the crease when in the right position. There is your swimming swan.

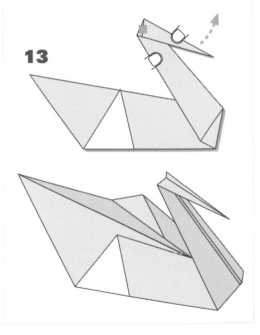

Useful items

DRINKING CUP

You will need

a square piece of paper about 20 cm x 20 cm. Use strong paper, such as copier paper, and the cup should hold cold liquid for several minutes without leaking.

1 Fold corner **A** up to the top. **2** Divide the fold line into twelve units. At about five units from the right, fold corner **B** across so that its upper edge is parallel to the base.

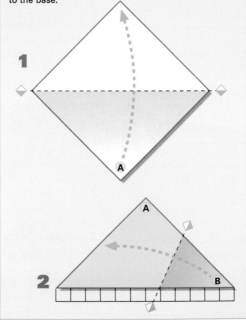

3 Fold down corner **A** over the top edge of the corner **B** flap. **4** Unfold it back up again.

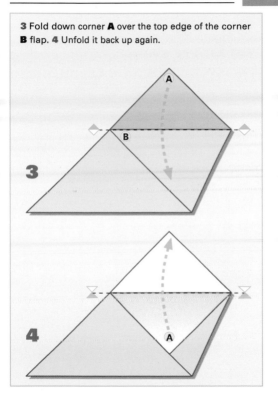

5 Tuck corner **A** inside the corner **B** flap pocket, pushing it as far down as it will go. **6** Flatten the top edge and turn the model over.

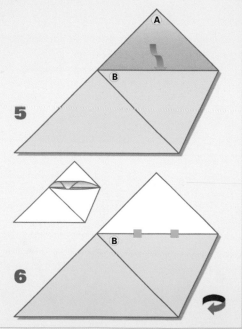

7 Fold corner **C** across to the opposite corner. **8** Fold down corner **D** along the top edge of the flap pocket made by **C**.

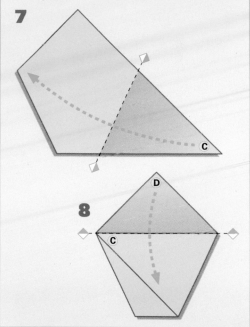

9 Fold corner **D** back up. **10** Tuck corner **D** into the flap pocket and flatten the fold. Carefully open the top central slit of the cup and pour in a cold drink.

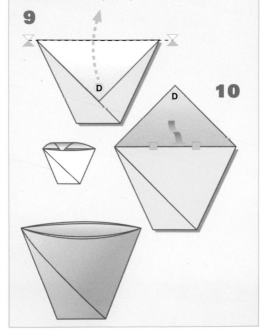

SUN HAT

You will need

a square piece of paper about 40 cm x 40 cm to make a sun hat to fit a child. A double page from a tabloid newspaper or a single page from a broadsheet cut square will do. The hat is a very good shade for children as the back flap protects the neck.

1 Fold corner **A** to the diagonally opposite corner. **2** Measure and mark the point halfway along the fold line. Now fold up corner **A** to that point.

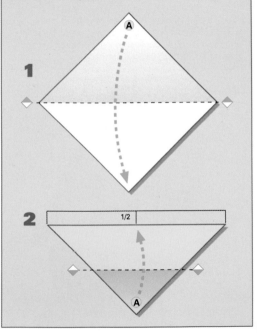

3 Fold in corner **B** to lie along the short fold line made in stage **2**. **4** Fold in corner **C** to do the same.

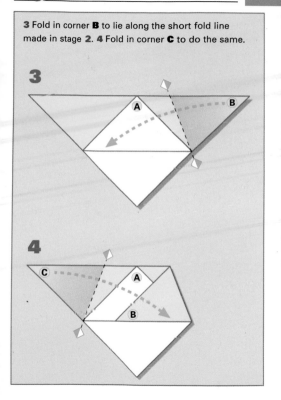

5 Unfold corners **B** and **C** and fold down corner **A** then **6** refold corners **B** and **C**.

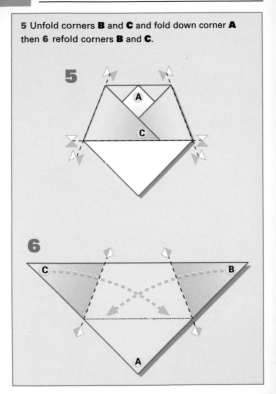

7 Fold corner **A** up so it tucks tightly into the pocket made by corner **C**. **8** Fold flap **D** up tightly against the bases of **B** and **C**.

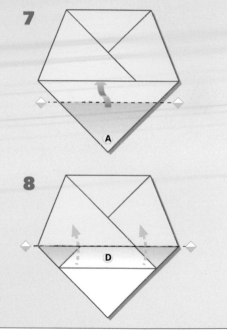

9 Holding down the back, open up the hat. Push the top down to make the 'ears'.

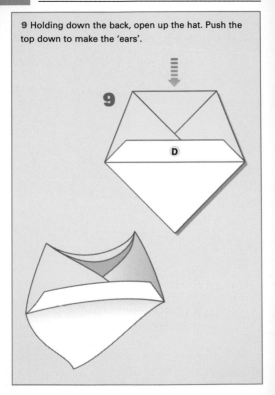

'PRINTER'S' HAT

You will need

a large sheet of newspaper. This traditional hat is made by printers using a double page from a broadsheet newspaper, measuring 74 cm x 58 cm.

1 Fold edge **A** to the bottom edge to halve the sheet.
2 Crease it vertically through the middle.

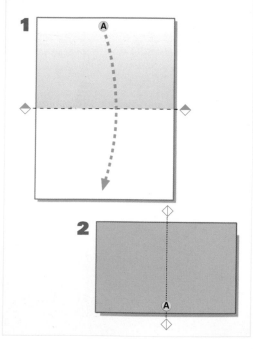

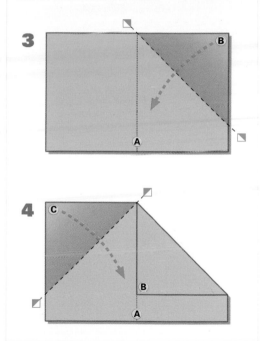

3 Fold down corner **B** to lie along the central crease line. **4** Fold in corner **C** to match.

5 Fold up flap **A** so the edge lies along the base of the triangular flaps. **6** Fold up flap **A** again along the base line to overlap the triangular flaps.

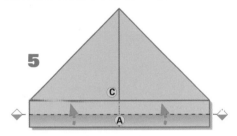

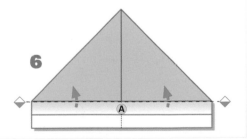

7 Turn the model over. **8** Fold in side **D** to meet the centre crease.

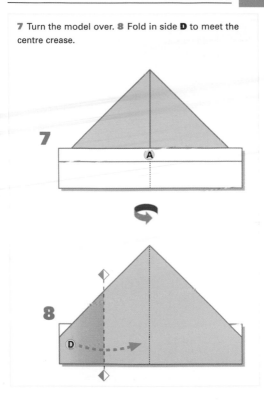

9 Fold in side **E** to the centre crease. **10** Fold up flap **F** so the edge meets the base of flap **A**.

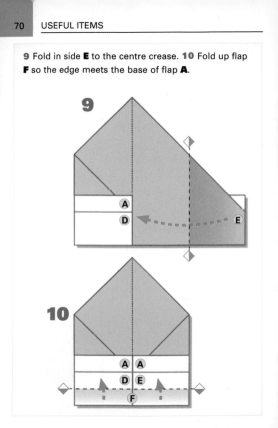

11 Fold up flap **F** again to overlap flap **A**. **12** Fold down corner **G**, tucking it behind flap **F**.

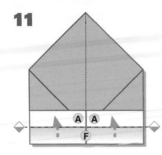

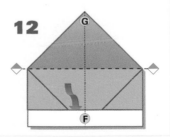

13 Turn the model through 90°. **14** Pull open the flaps **H** and **I**. Keep pulling them until points **J** and **K** come together.

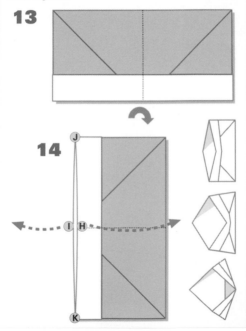

15 Fold down flap **L**, tucking it behind flap **J**. **16** Fold up flap **M**, tucking that behind flap **K**.

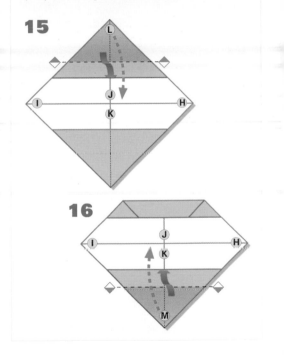

17 Pull open all the folds at **J** and **K** until you have a box-like shape, turn it over and there's your finished printer's hat.

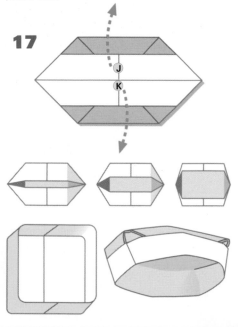

DECORATIVE BOX

You will need

a square piece of paper. You can use almost any size but a piece 20 cm square will make a box 5 cm wide. The box is shallower inside than the outside depth.

1 Measure and mark the centre point of the sheet.
2 Fold corner **A** in to touch the centre.

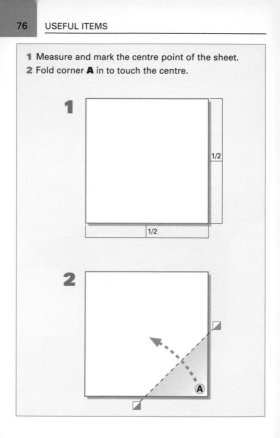

3 Now fold in **B** to touch the centre point. **4** Similarly corner **C**.

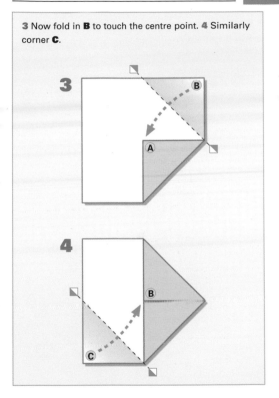

5 Now fold in corner **D** to meet the other three and **6** turn the model over.

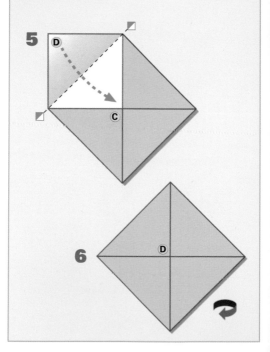

7 Fold corner **E** down to lie on corner **F**. **8** Fold corner **G** across to lie on **H**.

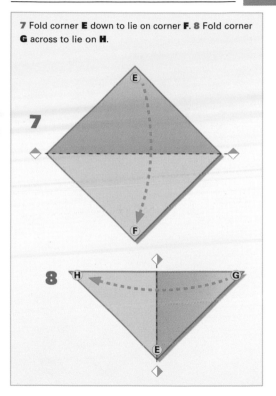

9 Pull only the top layer at **I** up and across, keeping the corner **E** in place and bringing **G** down. **10** Flatten the folds and turn the model over.

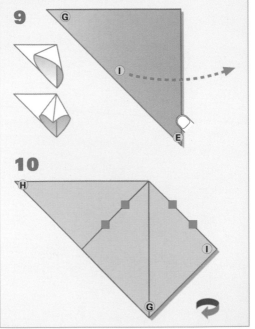

11 Similarly, pull **J** up and over to **I**, bringing **H** to the bottom. **12** Flatten the folds and rotate the model through 180°.

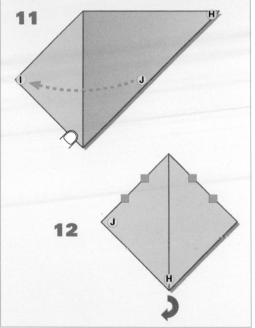

13 Crease the centre line by bringing the top layer of **H** down to the base. **14** Pull **K** and **L** up and out so that **H** is drawn to the base.

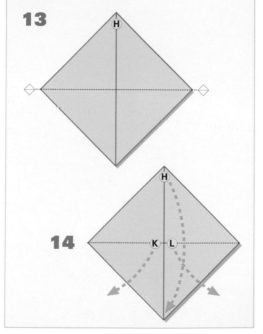

15 Flatten the creases and turn the model over.
16 Fold **G** down to the base and back up to make a creased centre line.

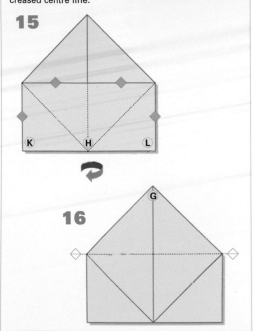

17 Pull **M** and **N** out from the centre, drawing **G** down to the base and flatten the folds. **18** Fold the top layer of **N** across to **M**. Now turn the model over.

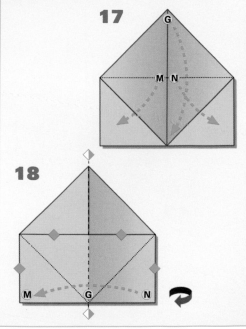

19 Fold **L** across to lie on **K**. **20** Fold the top layer of **M** into the centre.

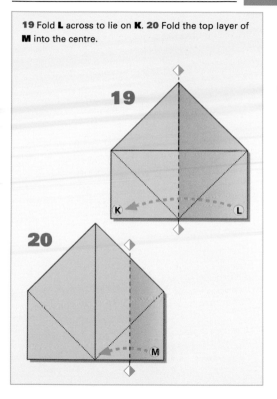

21 Fold the top layer of **L** into the centre to touch **M**.
20 Turn the model over.

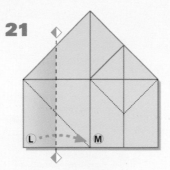

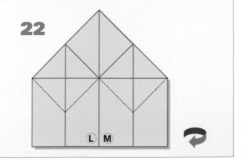

23 Fold **K** and **N** into the centre. **24** Fold the upper flap at **O** down to the bottom. **25** Turn the model over once more.

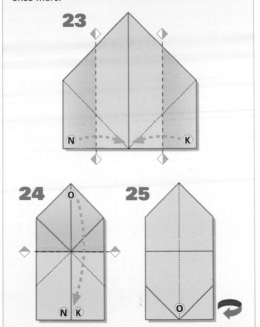

26 Fold flap **P** down to the bottom. **27** Pull flaps **O** and **P** out to the sides to open the box. You will need to push the base up and crease the edges.

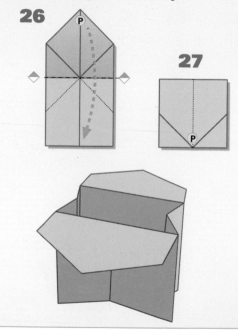

CONDIMENT OR SWEET DISH

You will need

a square piece of paper. A piece 20 cm square will make a dish that will hold after-dinner mints or small sweets. A smaller size would be suitable for a salt and pepper dish.

1 Crease the paper twice. **2** Fold corner **A** to the centre.

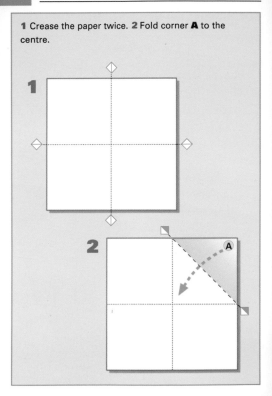

3 Fold corner **B** to the centre. **4** Fold corner **C** to the centre.

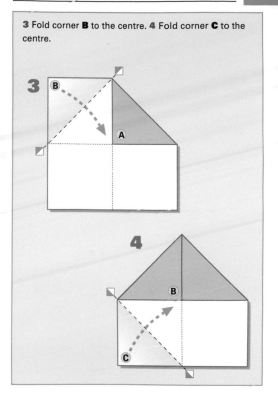

5 Fold corner **D** to the centre. **6** Turn the model over.

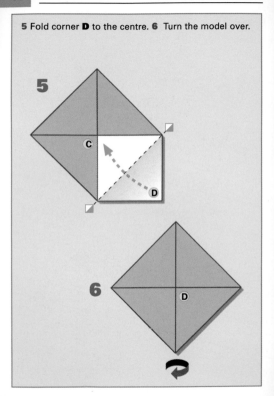

7 Fold corner **E** to the centre. **8** Fold corner **F** to the centre.

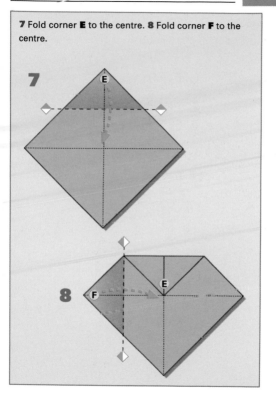

9 Fold corner **G** to the centre. **10** Fold corner **H** to the centre. **11** Firmly crease through the centre in both directions.

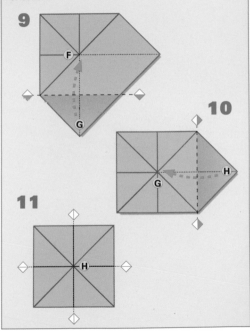

12 Fold **I** across the centre line. **13** Insert your index finger and thumb under the upper flaps **J** and **K**.

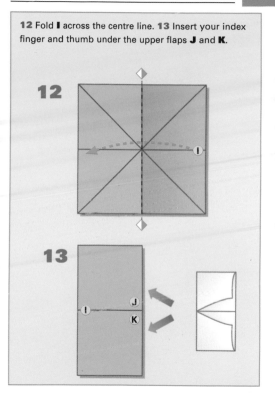

14 Carefully squeeze corners **L** and **M** together.
15 Turn the model over.

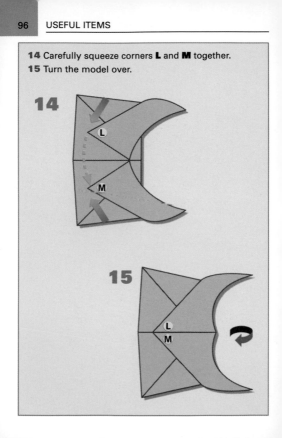

16 Insert fingers and thumb into **N** and **O**. Turn the model so it stands with the four compartments uppermost.

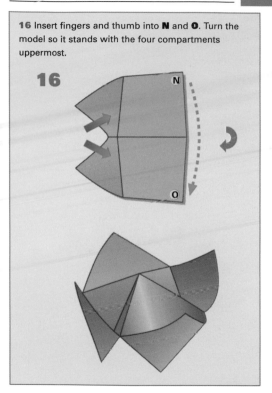

Decorative origami

FAN

> ### You will need
>
> an oblong piece of paper with the proportion 2:3. Don't make the fan too small as the folds will be very difficult to do and the fan will not be effective. Use gift wrapping paper or any other patterned paper for a decorative fan but the paper must be stiff enough.

1 Crease the middle of the sheet both vertically and horizontally **2**.

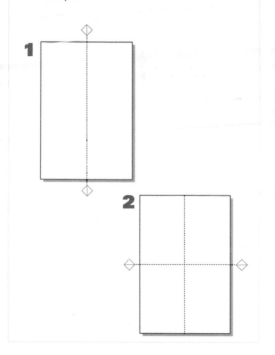

3 Use the shorter middle crease as a guide to make a third crease. **4** Using this as a guide, make a fourth crease. Turn the sheet round.

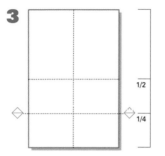

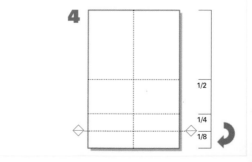

5 and 6 Repeat the creasing method for the other half of the sheet.

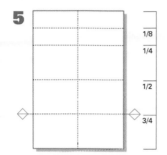

5

1/8
1/4
1/2
3/4

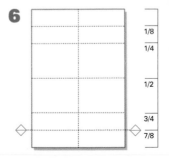

6

1/8
1/4
1/2
3/4
7/8

7 Fold then unfold **A** so that the bottom edge touches the 1/4 crease. Turn the sheet round. **8** Fold and unfold edge **B** to the 3/4 crease. Turn the sheet over.

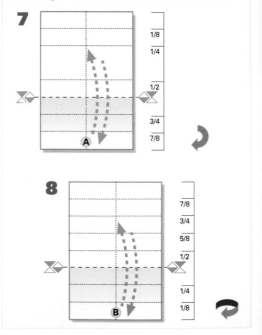

9 Use the first crease line as a guide and fold edge **B** up to that line. 10 Fold flap **B** and the underlying portion of the sheet back under.

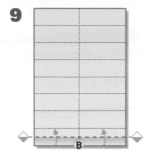

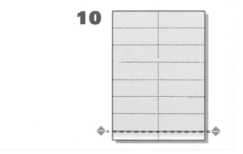

11 Fold edge **C** and the underlying flap up to the next crease line. **12** Fold back under and repeat to concertina fold the whole sheet. Rotate the model 90°.

11

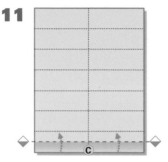

12

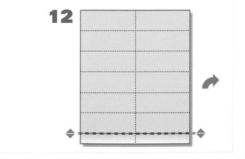

13 Fold over just the top layer at corner **D**. **14** Fold back the top layer of corner **E**. **15** Crease the next layer, **F** and **G**, along the new fold lines.

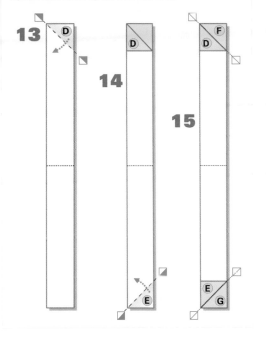

16 Tuck **F** back within itself to lie under corner **D**. Tuck **G** inside and under **E** then repeat steps **15** and **16** with all other corners except the last two.

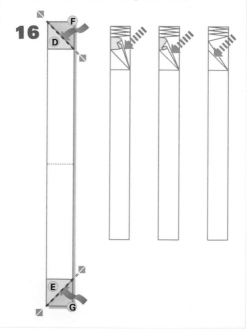

17 Fold corners **T** and **U** underneath. **18** Unfold the top layer **V**. **19** Fold the concertina in half along the original crease line.

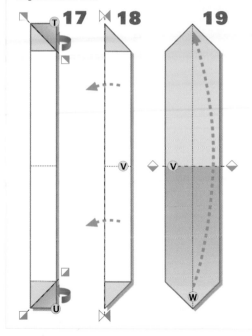

20 Fold up corner **X**. 21 Carefully tuck the outer flap **V** inside the adjoining pocket in the concertina 22. 23 Holding the central three segments tightly together at the back, pull down the bottom flap, **Y**.

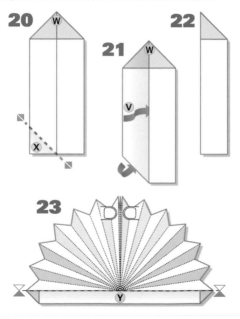

24 Fold flap **Y** up and tuck inside the adjoining flap to complete the fan.

24

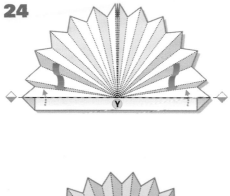

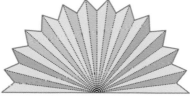

BUTTERFLY

You will need

a square piece of paper, about 20 cm x 20 cm, to make a butterfly with a wing span of approximately 20 cm. Use patterned papers to make colourful butterflies or plain paper and add patterns to the finished model.

1 Crease the horizontal, vertical and diagonal centre lines as shown. **2** Fold corner **A** to the diagonally opposite corner.

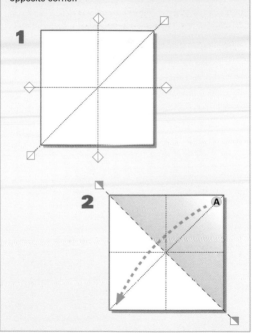

3 Holding the centre of the base of the model, pull corner **A** up, drawing **B** across to the opposite side and **C** down to sit on **A**. **4** Now pull **C** back across, drawing **B** down to the centre base line.

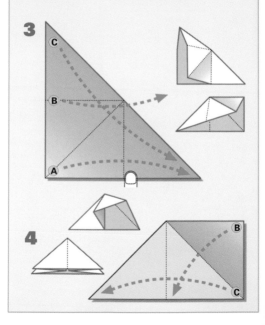

5 Rotate the model 180°. **6** Measure the height and fold slightly above the halfway mark so that **D** extends about 5 mm beyond the top edge at **B**.

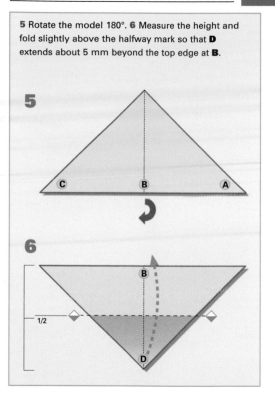

7 Fold the top layer of **E** down along the fold edge.
8 Similarly, fold the top layer of **F** down along the fold edge.

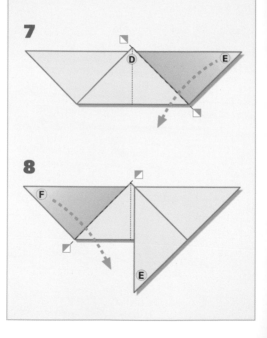

9 Fold **G** across to the opposite side. **10** Unfold the model to make a crease.

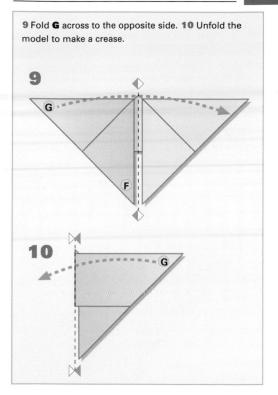

11 Turn the model over. **12** Fold **H** across a line running up from the edge of the lower wing.

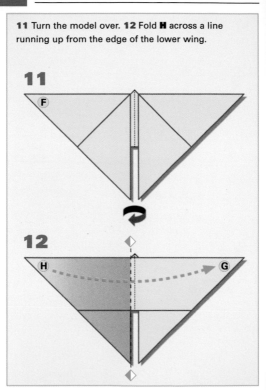

13 Unfold **H**. **14** Now fold **I** across using the edge of the lower wing as a guide.

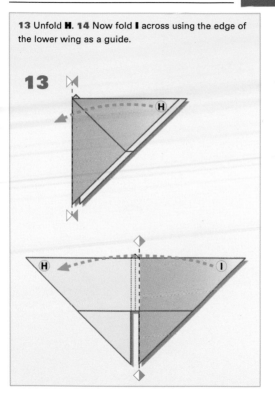

15 Unfold ▌ and even the folds to display the butterfly.

15

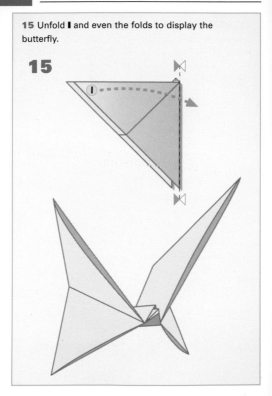

FLOWER

You will need

a piece of paper between 15 cm and 25 cm square to make a lifesize flower. Patterned paper or foil paper will make attractive imitation blossom.

1 Fold the sheet in half. **2** Using another sheet of the same size, mark a point on the fold edge one sheet length from the opposite bottom corner.

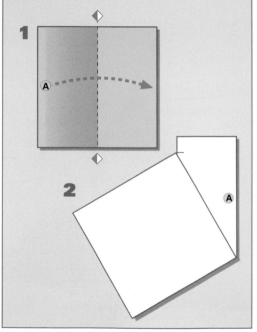

3 Fold the top layer of corner **B** across to lie on the mark. **4** Flatten the fold line and cut along the edge. **5** Turn the resulting triangle over.

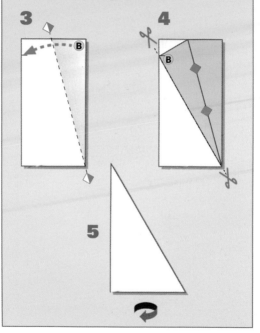

6 Fold **C** down to the opposite corner. **7** Turn the sheet over.

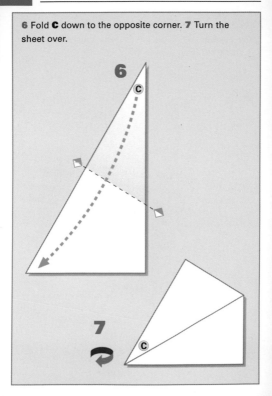

8 Fold **D** down to the base. 9 Fold **D** again up to the new fold line. 10 Unfold all folds and rotate so that the apex is pointing down.

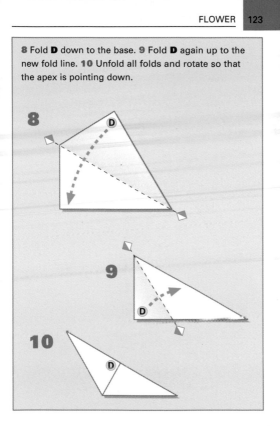

11 Fold **F** up across the centre of the triangle.

12 Holding the right-hand side of **F** flat, pull **G** across the mid-vertical line.

11

F

12

F

G

13 Keeping **G** flat, pull **H** across to sit on **G**. **14** Turn the model over.

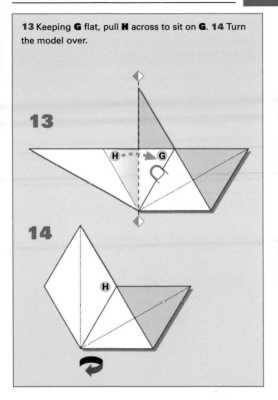

15 Keeping **I** flat, pull **J** across to the opposite corner. Make sure that the flap behind **K** stays tucked in.

16 Fold the upper flaps of **J** and **K** to the centre.

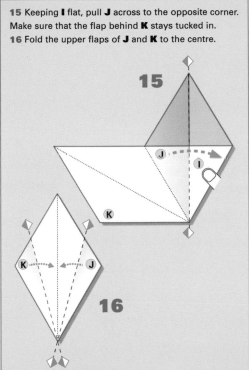

17 Turn the model over. **18** Fold the upper flaps **L** and **M** to the centre.

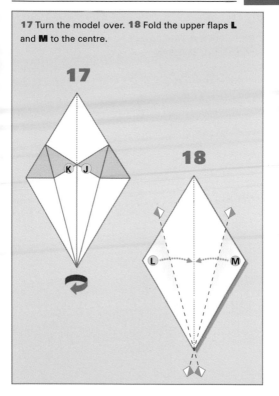

19 Fold the upper flap of **G** across the mid-vertical line. **20** Fold **G** and **N** to the centre.

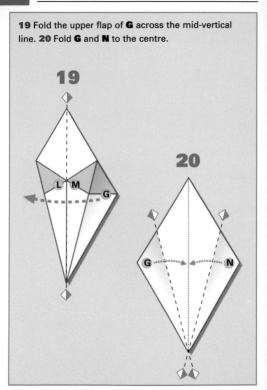

21 Fold the upper flap **F** down to the bottom point.
22 Turn the model over.

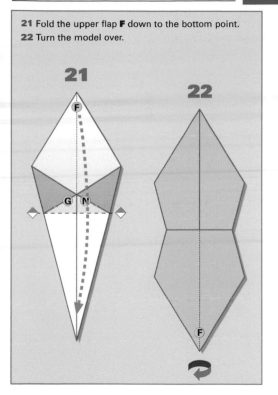

23 Fold **O** down to the bottom. **24** Fold **P** across to open up the third petal.

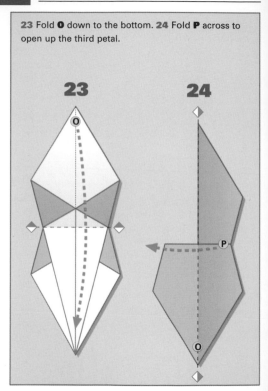

25 Fold **Q** to the bottom. **26** Lift up the petals to finish the flower.

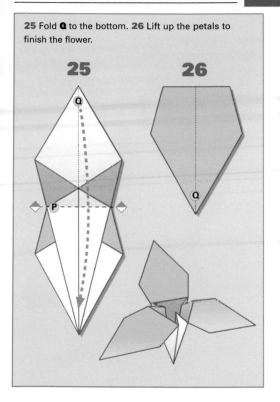

STAR

> ### You will need
>
> two square pieces of paper about 20 cm x 20 cm. Use different colours, or even different textures, to add interest. Foil papers will also create a decorative effect.

1 Fold one sheet in half. **2** Use the second sheet to mark the fold line at a point one sheet length from the bottom outer corner. Fold and mark the second sheet in the same way

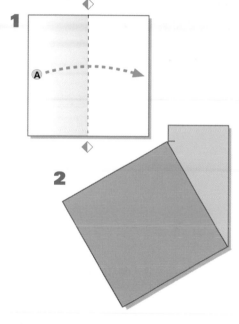

3 Using the first sheet, fold corner **B** across to touch the mark. **4** Flatten the crease and cut along the edge. **5** Unfold the triangle.

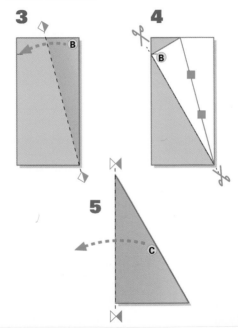

6 Fold **D** down to sit on **E**. **7** Unfold **D**.

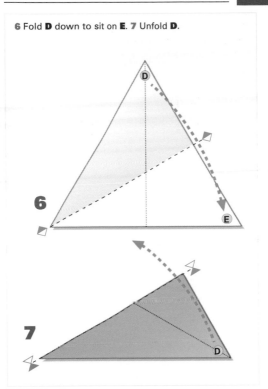

8 Fold **D** down to sit on **F**. 9 Unfold **D**.

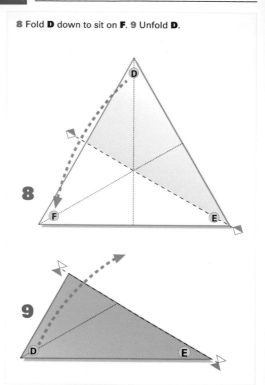

10 Fold **D** down to touch the centre point. **11** Fold **E** into the centre point.

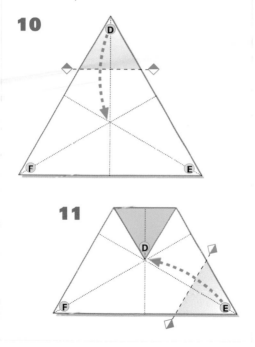

12 Fold **F** into the centre point. 13 Turn the model over. 14 Make sure that flaps **D**, **E** and **F** stay in position.

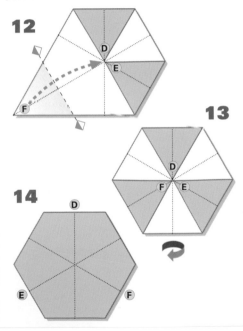

15 Fold **G** into the centre so that part of flap **D** and part of flap **E** is folded. **16** Fold **H** into the centre. **17** Fold **I** into the centre.

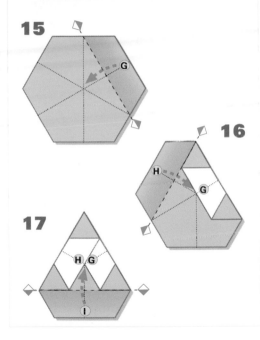

18 Holding parts of flaps **H** and **I** in place, pull out the lower part of flap **G** and tuck flap **I** underneath it.

19 Turn the model over and follow steps **4** to **19** to make a second triangle.

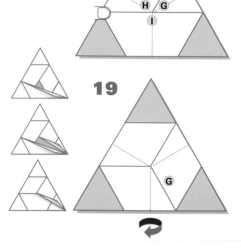

20 Rotate one triangle so its apex is down. **21** Slide
the one triangle under the top flap of the second one.

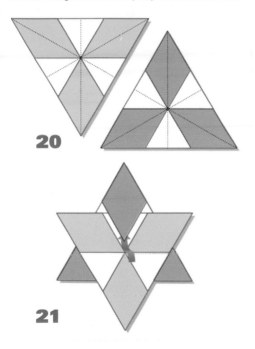

22 Carefully lift each of the flaps of the rearmost triangle and pull them over the front one to make the star.

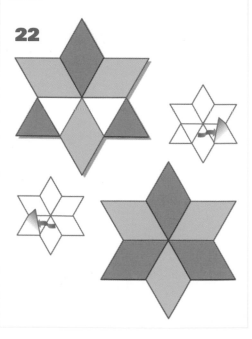

22

Paper toys

BIRD

You will need

a piece of paper about 20 cm square. Use paper with a small bright pattern to make an exotic bird.

1 Crease the diagonal. **2** Fold **A** so its edge lies along the creased diagonal line.

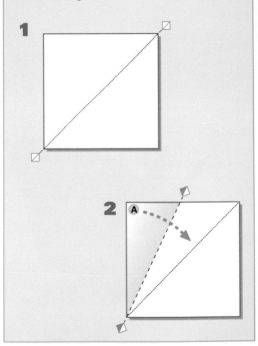

3 Fold **B** also to lie along the diagonal crease. **4** Turn the model over.

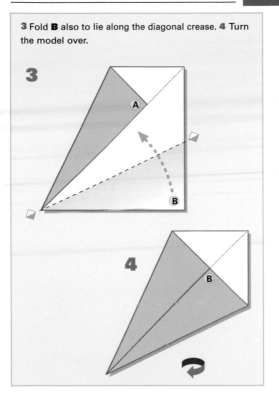

5 Fold **C** to lie on the centre crease. 6 Turn the model over.

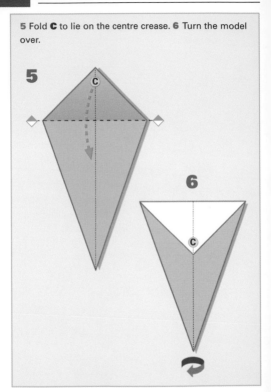

7 Fold **D** and **E** to the centre line. **8** Use a ruler to make a crease from one corner to the centre line.

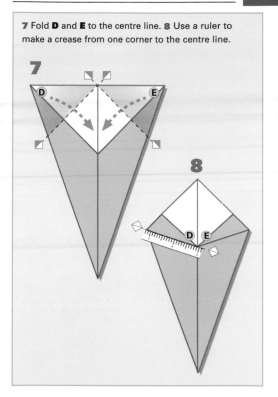

9 Similarly crease from the other corner to the centre line. **10** Unfold **D** and **E** to the centre line.

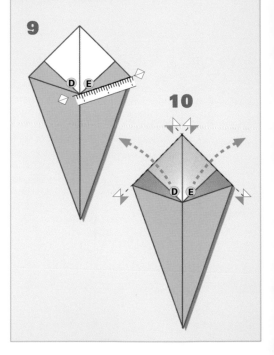

11 Holding the centre of the model, lift the top layer of corner **F** across so that **D** is pulled down to the centre. **12** Now pull the top layer of **G** down so that **E** is drawn to the centre.

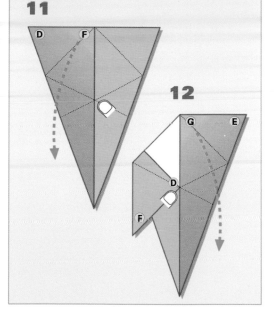

13 Measure and mark points 1/4 along the left edge of **F** and half way down the right edge. **14** Fold **F** across a line between the marked points.

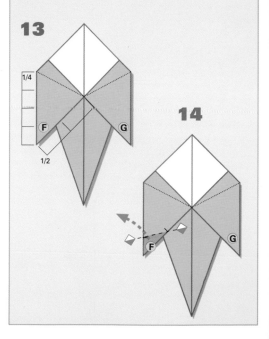

15 Measure and mark points 1/4 along the right edge of **G** and half way down the left edge. **16** Fold **G** across a line between the marked points.

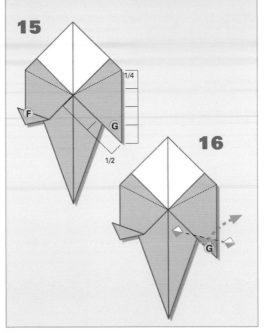

17 Fold the model in half across the centre line.
18 Measure and mark points half way down the left edge and 1/4 way down the right egde.

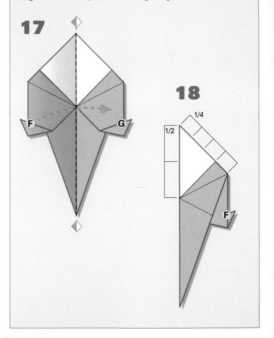

19 Crease a line between the marks. **20** Holding the model together, push inwards corner **H** and flatten the folds.

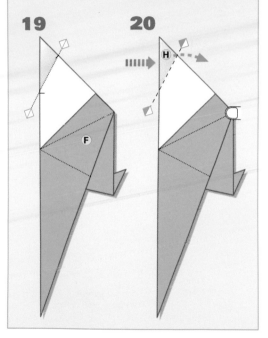

21 Cut a straight line up to form the tail. **22** Fold up wing tip **I** and turn the model over.

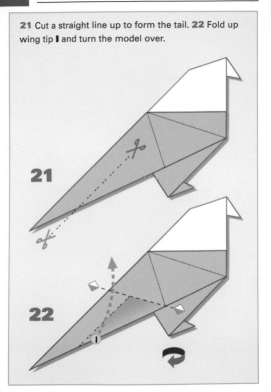

23 Fold up wing tip **J** to match **I**. The completed model should stand on its feet.

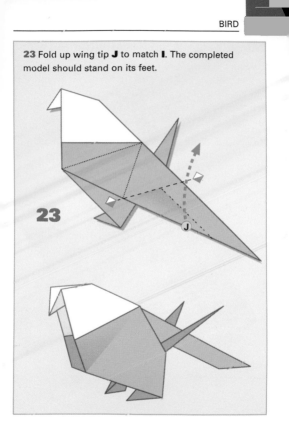

23

SAMURAI (JAPANESE WARRIOR) HELMET

You will need

a square piece of paper 52 cm x 52 cm to make a
helmet to fit a young child. Use a fairly thin paper
otherwise the helmet will slip. Japanese warriors –
samurai – wore metal helmets which looked like this.

1 Fold **A** diagonally across the centre of the sheet.
2 Fold corner **B** down to lie on **A**.

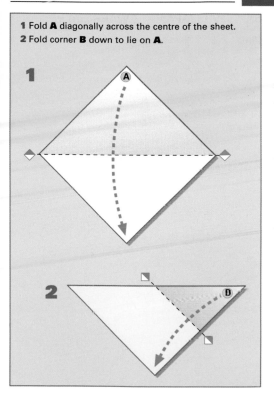

3 Fold corner **C** to lie on **A**. **4** Fold **B** and **C** up to the top corner of the model.

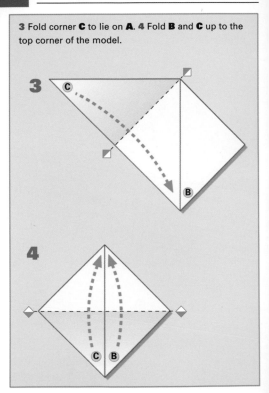

5 Fold **B** out to the side. The angle for this is not critical. **6** Fold **C** out to the other side to mirror **B**.

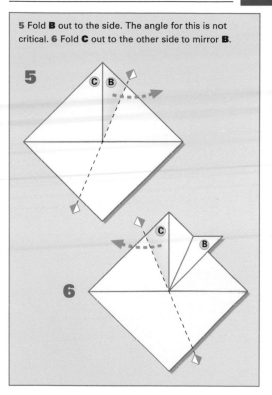

7 Measure points about one fifth down each of the lower sides and fold **A** up across a line between those points. **8** Fold flap **D** up across the centre line.

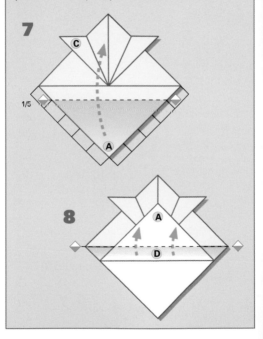

9 Turn the model over. **10** Measure and mark a point about one seventh in from either side. Fold **E** in at the right-hand mark.

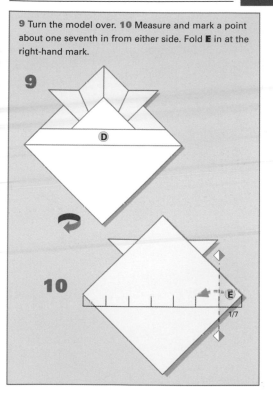

11 Similarly, fold **F** in at the left-hand mark. **12** Fold **G** up to the top.

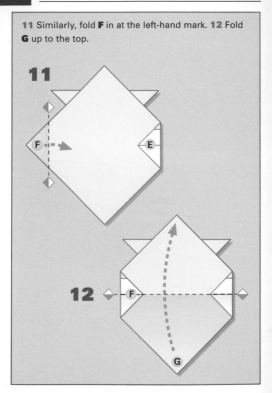

13 Turn the model over and open the gap at the base to make the helmet.

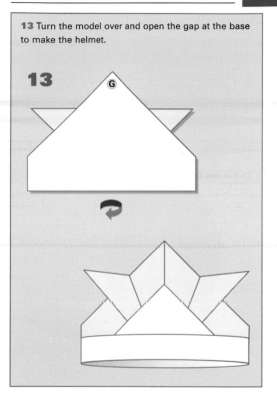

DRAGONFLY

You will need

a piece of paper about 20 cm square. Dragonflies are usually brightly coloured. Use foil paper or delicate but colourful patterned paper.

1 Fold corner **A** across the diagonal. **2** Fold corner **B** to the opposite corner.

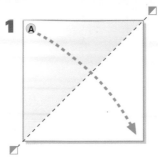

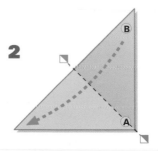

3 Pull the top layer of **C** up and across to pull **B** down to the bottom right corner. **4** Flatten the folds and turn the model over.

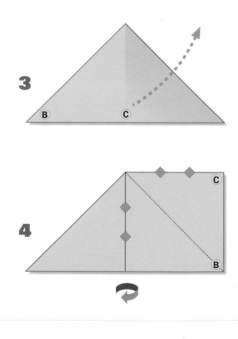

5 Pull the top layer at **D** up and across to bring **E** to the bottom left. 6 Flatten the folds and rotate the model so corner **E** is at the base.

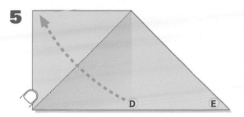

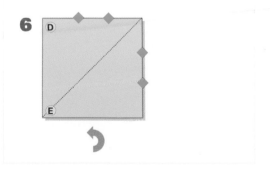

7 Fold in the top layer of **D** to the centre crease.
8 Fold in the top layer of **F** to the centre crease.

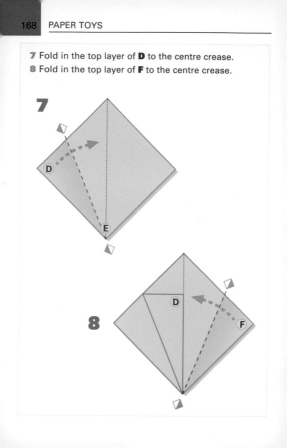

9 Turn the model over. **10** Fold in **G** to the centre crease.

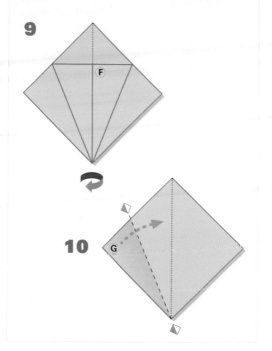

11 Fold in **H** to the centre crease. **12** Fold down flap **I**.

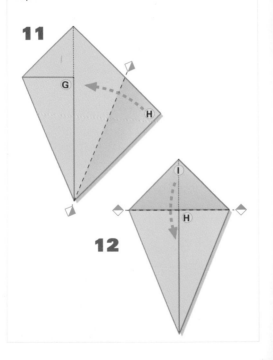

13 Unfold flap **I**. **14** Unfold flaps **G** and **H**.

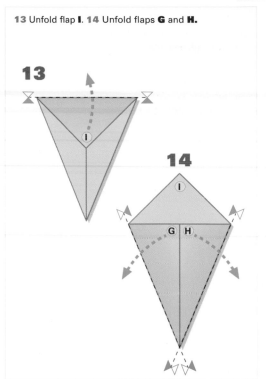

15 Lift the top layer of corner **J** and pull it up, bringing the edges **G** and **H** to meet in the centre.
16 Turn the model over.

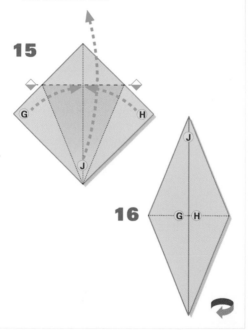

17 Unfold flaps **D** and **F**. **18** Pull top layer of corner **E** up to the top, drawing edges **D** and **F** to the centre.

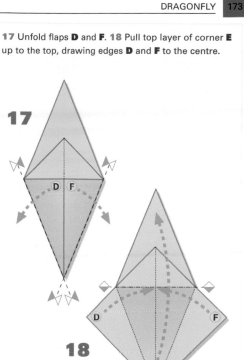

19 Fold in corner **K** to the centre line. **20** Similarly, fold in corner **L** to the centre line.

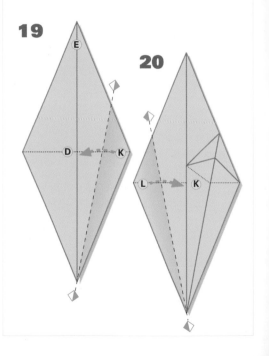

21 Turn the model over. **22** Now fold in corner **M** to the centre.

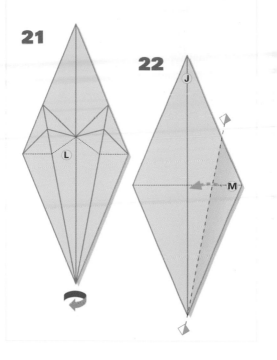

23 Fold in corner **N** to the centre line. **24** Hold the model firmly together and pull **O** out to the side. This will reverse the inner fold of the segment.

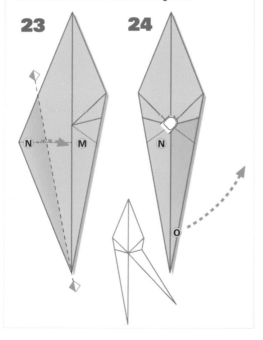

25 Flatten the crease and turn the model over.

26 Now pull segment **P** up to the side, reversing the inner fold.

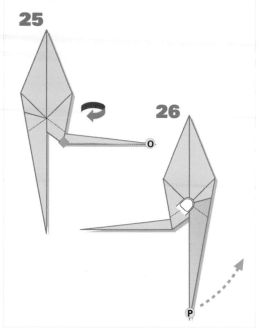

27 Fold the end third of **P** back on itself. **28** Fold most of **P** back again. **29** Fold **P** back under to tuck inside the main segment. **30** Flatten the folds.

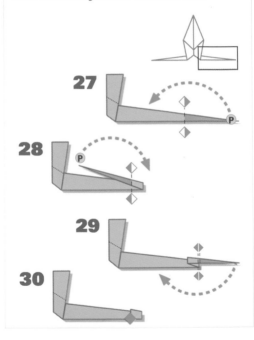

31 Complete the model by pulling the wings out to either side of the body so that the top surface **Q** is flattened.

31

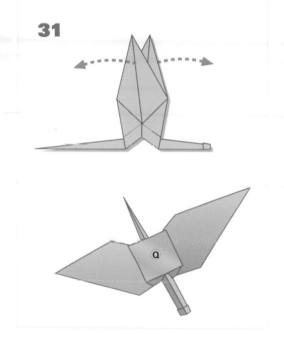

Stationery items

LETTER KNOT

> ### You will need
>
> a piece of paper 15 cm x 30 cm for the method shown.
> With practise you can use any size as long as it is at
> least twice as long as it is wide. Use a light-coloured
> paper and write your message on the side that is to be
> folded in. The name of the recipient can be written on
> the outside of the finished piece.

1 Crease the paper. **2** Fold edge **A** to the centre crease.

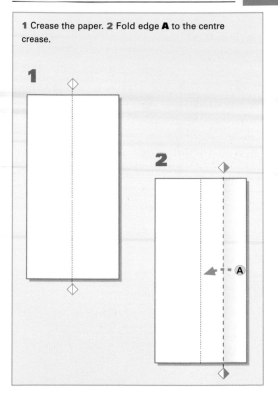

3 Fold **B** to the centre crease. 4 Fold **A** to lie on **B**.
5 Lightly mark points on either side 140 mm from each end.

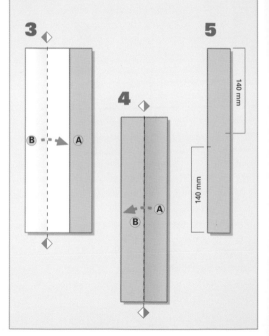

6 Use a ruler to help make a crease between the two marks. **7** Fold **C** up and across. **8** The upper edge of **C** should lie slightly to the right of the mark and extend about 2 cm beyond. Turn the model round.

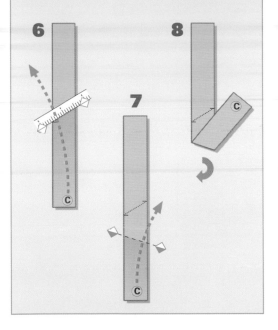

9 Fold **D** up and across as for **C** in step **7**. 10 Unfold **D**. 11 Fold **D** across the central crease line. 12 Reverse fold **D** back under the model.

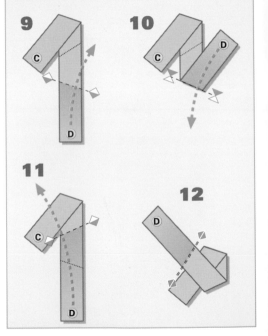

13 Tuck **D** through the flattened loop, passing behind **C** but over the central part of the model.

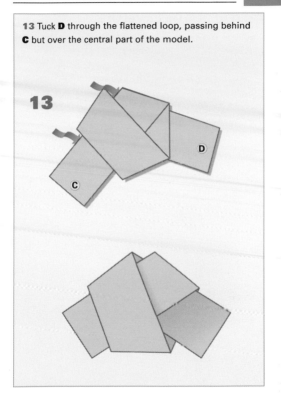

LETTERFOLD OR ENVELOPE

You will need

a square piece of paper. The size will depend on the size of envelope you require. Use a fairly stiff paper but not too thick. You may need to stick an address label on paper that has a strong pattern or dark colour.

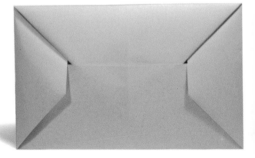

1 Crease the paper vertically. **2** Measure and mark the centre point.

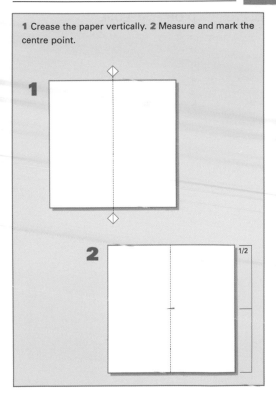

3 Fold **A** up to touch the centre mark. **4** Mark points 1/6 in from each edge.

3

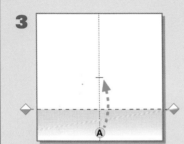

4

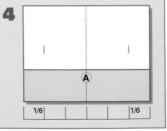

5 Fold **B** across the lef-thand mark. **6** Fold **C** across the right-hand mark.

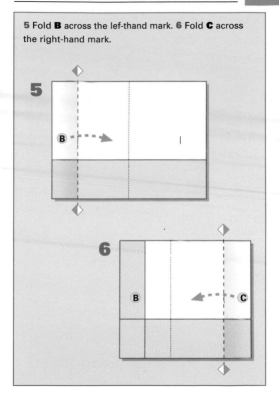

7 Unfold flaps **B** and **C**. **8** Fold in corners **D** and **E**.

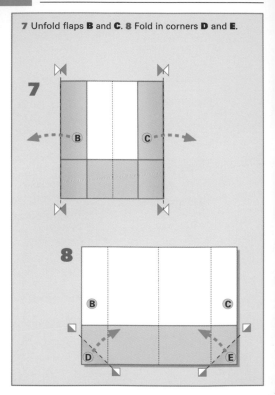

9 Refold flaps **B** and **C**. 10 Fold down corners **F** and **G** to the centre crease.

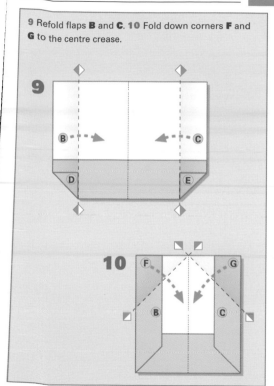

11 Fold down flap **H** and tuck under **I**.

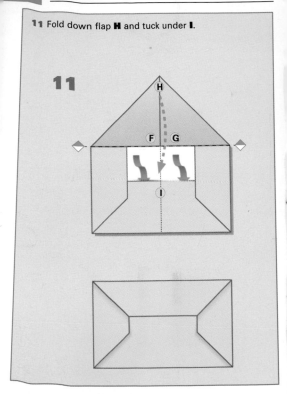